To our farm friends who love our animals as much as we do remember to work hard, be kind, and dream big.

-Simply Country Ranch
Ben & Lauren Denny

Published by PrintNinja
in 2020
Printed in PRC

(Paperback)
ISBN no.: 978-0-578-77667-5

Written by Ben & Lauren Denny
Simply Country Ranch

Illustrated by Marina Halak

Winston

and the Missing Apples

"It's harvesting day! Come on Farmer Ben,
let's go to the orchard!"

Winston's favorite part of harvesting day is when Farmer Ben shares the leftover apples with all the friends on the farm.

Winston and Farmer Ben enter the apple orchard.
Someone has been eating the apples!

"Winston, if someone keeps eating the apples, I don't know if there will be apples for everyone to share. Can you help me solve the case of the missing apples?"

Winston thinks, "Where should I look first, the farm is so big? The apples could be anywhere."

Mr. Rupert and the sheep call out:
'Good morning, Winston! We are headed to the pasture, would you like to come with us?"
"I can't this morning guys, I have to find the missing apples," Winston replies.
"Do you know anything about missing apples, Rupert?"
"No, I don't, but I bet the chickens do. They are always in the orchard."

The chickens were in the orchard just as Rupert said they would be.
"Hey ladies, have you seen
anyone in the orchard eating apples?"
"Sorry, Winston, we haven't but we will help you look."

"Winston! Winston! Come quick," Ethel cries, "I found an apple!" Winston thinks, "Hmmm, how did this apple get all the way over here by the big willow tree?"

"I found another! It looks like a trail leading into the willow tree," cried Harriet.

"Come on ladies, let's see where the trail leads!"
"Oh, no! You're crazy if you think we are going in there," Dorothy says.

Ethel warns, "Be careful, Winston." Martha covers her eyes, "I can't watch."

"Well, if you're too scared, I'll go in myself,
Farmer Ben is counting on me."

Winston musters up all his courage and steps under the willow tree.
Martha turns her back and whispers, "He's so brave."

Several minutes pass in silence and the chickens begin to worry.
"It's been too long."
"Should someone get Farmer Ben?"
"What are we going to do?"

Just then Mr. Rupert and the sheep come over and say, “What’s going on?”
Martha shrieks,“It’s Winston! He went under the willow tree!”
Fanny gasps, “Doesn’t he know a monster lives in there?”

Just then, a loud crash comes deep within the willow tree.

in there!

Winston races to find Farmer Ben. “Come quick! It’s a monster! A monster has been taking the apples. It’s hiding under the willow tree.” Farmer Ben drops his buckets and heads straight for the tree.

Farmer Ben steps into the willow tree.
A few minutes later Farmer Ben calls out,
"There's no monster! It's just Ruby."

Ruby peaks her head out of the branches.
Winston says, “Ruby, you’re not supposed to be in the orchard. Let’s get you back with the rest of the fold.”

As Winston and Ruby are walking back, they see all the Scottish Highland Coos waiting at the fence for harvesting day.

Darla huffs, “Why not?” Ruby whispers, “I am sorry, but I ate the apples.”

Lola begins to cry, "I was really looking forward to eating apples." Maggie states, "That's okay, we will get apples next year." The fold walks away.

"I really love apples, but I didn't think about leaving enough for everyone else.
What should I do?" Ruby asks Winston.

"It is important to think of others.
I'll ask Farmer Ben if we can share some of the apples he is taking to the market. I'll be right back!"

Moments later, Farmer Ben and Winston return with a bucket of apples.

Ruby says to Farmer Ben, “Thank you for sharing your apples.
I learned that it is important to put others first
so we can all have a few apples to share.”

Just then, the fold runs back to the fence and everyone cheers as Farmer Ben and Winston pass out apples.

"Thank you, Farmer Ben, for sharing your apples with the friends on the farm."
Farmer Ben looks at Winston, "We had quite the harvesting day.
You were so helpful solving the case of the missing apples. I think we all learned how important sharing is."
Winston says, "We sure did!"

Marina Halak

Marina is a freelance illustrator based in Germany.
Inspired by her own childhood, children, nature, magical moments, sea and fairytales.
If she doesn't draw (rarely), she enjoys spending time with her boyfriend,
tries to teach her dog to speak, collects picture books and magical, and beautiful
things, cooks and reads books.

You can learn more about Marina at www.grosseaugenart.com.

Simply Country Ranch | Ben & Lauren Denny

The characters in the book live on their ranch in Oklahoma. They are excited to share a piece of their ranch living as it comes to life in their book. Ben worked in construction and Lauren was an elementary school teacher before they decided to ranch full time.The simple country living is a passion of theirs and they love that each animal truly has their own personality.

You can learn more about their story at www.simplycountryranch.com